How to Make a Photocatalytic Air Purifier
by Stefan Hollos and J. Richard Hollos

Abrazol Publishing
an imprint of Exstrom Laboratories LLC
662 Nelson Park Drive, Longmont, CO 80503-7674 U.S.A.

Contents

Preface

We write books, so we spend a lot of time indoors. When we go outside the air seems so much fresher than inside. Is there anyway to recreate that fresh air feeling indoors? That's a question we've been trying to answer for a few years now. We think we've found an answer. It's called photocatalytic air purification. What exactly this means is described in the next section. The final section of this guide describes in detail how to actually make an air purifier that uses photocatalysis to remove volatile organic chemicals (VOCs) from the air.

The photocatalytic air purifier which we've created and describe below has made a positive difference to us. We set it on a table next to where we work and run it whenever we're working. When it's running, the air seems fresh, similar to having a nearby open window. It's not magic but it has reduced the incidence of itchy watery eyes, irritated throat, runny nose and general feeling of malaise that VOCs can produce.

What is Photocatalytic Air Purification?

To answer this question let's begin by looking at air impurities. Ideally air should consist of mostly nitrogen N_2, oxygen O_2, argon Ar, and water vapor H_2O.

But air contains trace amounts of many other gases. Most of these fall into the category of volatile organic compounds or VOCs. Some VOCs are of natural origin but many are human-made.

Carbon dioxide CO_2 and methane CH_4 are two of the major VOCs found in the atmosphere. They originate from natural processes as well as human activities. At low concentrations they do not cause health problems. There are many other VOCs generated by plants and animals that do not cause health problems at the levels commonly found in the natural environment.

Some human-made VOCs on the other hand can potentially cause serious health problems. This is especially true in an indoor environment with poor ventilation where concentration levels can become quite high. There are countless sources of VOCs in an indoor environment. Building materials, furniture, carpets, paints, glues, and sprays are common sources. They can produce VOCs such as:

- Formaldehyde

- Acetaldehyde

- Acetone

- Benzene

- Toluene

- Butane

Some of these are known carcinogens. They can also cause damage to the central nervous system as well as the liver and kidneys. Symptoms of exposure to these VOCs can include headache, dizziness, mental confusion and eye, nose and throat irritation. If you experience these symptoms on a regular basis it could be due to high VOC exposure in your work or home environment.

The purpose of photocatalytic air purification is to reduce the concentration of VOCs in the air. As the name implies, it does this using light in combination with a catalyst. A catalyst is a substance that initiates or speeds up a chemical reaction but is not itself consumed or transformed in the reaction. By far, the most common photocatalyst used is titanium dioxide TiO_2. It is cheap, nontoxic and readily available.

Titanium dioxide is usually used as a powder. The powder is mixed with a binder to form a paint that is deposited on a surface (substrate). The particles that make up the powder are small crystals of average diameter in the 10 to 100 nanometer range. The crystals come primarily in one of three forms called anatase, rutile and brookite. A small percentage of the particles are not in crystal form and are said to be amorphous. The type of crystal is important because it has an affect on the photocatalytic properties.

As an example, one of the most common TiO_2 powders used in photocatalysis is called Aeroxide® P25. One study (ref. 36) of this powder showed it to be 76% anatase, 10.6% rutile and 13% amorphous particles. Another study (ref. 37) found the percentages to be 78% anatase, 14% rutile and 8% amorphous so there is some variation. The producer of the powder specifies the average particle size to be 21 nm.

TiO_2 crystals are semiconductors. This means the crystal has a band of energies called the valence band separated from another band of higher energies called the conduction band as shown in figure 1 for the anatase form of TiO_2. In the lowest energy state of the crystal, all the electrons have energies at or below the top of the valence band.

Electrons with energies in the valence band are bound to atoms in the crystal lattice. Electrons with energies in the higher energy conduction band are not bound and can move through the lattice. The energy difference between the top of the valence band and the bottom of the conduction band is called the band gap. This is the

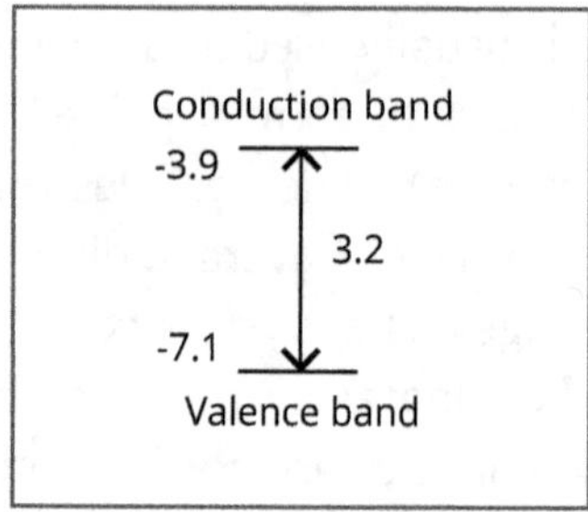

Figure 1: Energy band structure in TiO_2 in units of eV (electron volts).

minimum energy required to move an electron from the valence band to the conduction band.

The photocatalytic process starts when a photon with an energy larger than the band gap is absorbed by one of the TiO_2 particles. This moves an electron up to the conduction band where it can now move through the lattice. The absence of that electron from the valence band acts like a positive charge called a hole. The hole can also move through the lattice.

Two things can now happen. The electron can fall back down into the valence band by releasing energy in the form of heat or light (photon emission), annihilating the hole in the process. This is called recombination and it does not lead to photocatalysis. Alternatively the electron and hole can migrate to the crystal surface where they can interact with molecules adsorbed onto the surface. This interaction can change and break down those molecules.

The most common molecules with which the electrons and holes interact on the surface are oxygen O_2 and water

H_2O. The hole in the valence band can take an electron from a water molecule on the surface, creating a hydroxyl radical OH and an ionized hydrogen atom H^+ (a proton). This process is called water oxidation.

The electron in the conduction band can be taken up by an oxygen molecule on the surface of the crystal, creating what is called a superoxide ion, O_2^-. It can also be taken up by the ionized hydrogen atoms created in the oxidation of water by holes in the valence band, creating hydrogen gas H_2. These processes are called the reduction of oxygen and hydrogen respectively.

The products produced in the oxidation and reduction process described above, especially the hydroxyl radical, are then able to break down VOCs in the air into harmless molecules. The hydroxyl radicals can turn many VOCs into harmless carbon dioxide CO_2 and water H_2O. The superoxide ions and hydroxyl radicals can also inactivate (kill) microorganisms in the air.

What we have just described is photocatalytic air purification in the simplest possible terms. Now let's look at a few of the details.

Since it all begins with light, we should start by looking at exactly what kind of light is needed to get the process going. Light, being an electromagnetic wave, comes in many wavelengths and each wavelength has its own energy. Light with wavelength λ has energy

$$E = \frac{hc}{\lambda}$$

where $h = 4.1356679 \cdot 10^{-15} \text{eV} \cdot \text{s}$ is Planck's constant and $c = 2.99792458 \cdot 10^{8} \text{m/s}$ is the speed of light in a vacuum.

When light interacts with matter it behaves as though it exists in the form of discrete particles, called photons, with each photon having exactly this energy. In order for an electron in the valence band of TiO_2 to gain enough energy to make it up to the conduction band, it must absorb a photon with an energy equal to or greater than the band gap. The band gap energies and corresponding wavelengths for the three most common forms of TiO_2 are shown in the following table.

Form	Energy(eV)	Wavelength(nm)
Anatase	3.2	387
Rutile	3.0	413
Brookite	3.3	376

Note that there is some variation in these values in the scientific literature. This is probably due to the fact that band gap energies are somewhat dependent on the size of the particle as well as what the particle is in contact with. The important point is that the band gap energies are such that the TiO_2 particles only absorb light in the violet to ultraviolet wavelength range, 380 to 420 nanometers. There is very little of this wavelength of light in common indoor lighting or in light from the sun.

A lot of work in the field of photocatalysis has been done to try to reduce the band gap energies of TiO_2 so that a greater range of indoor light and light from the sun can be used. The wavelengths of light that the human eye can detect ranges from about 400 to 700 nanometers. The wavelength spectrum of indoor light is of course also in this range as is much of the light from the sun. Since human biology is well adapted to this wavelength range it would be most convenient and efficient to have a photocatalyst that operates over most of the range.

There has been some success in lowering the band gap energy of TiO_2 by doping it with various compounds, nitrogen compounds for example, or by mixing it with things like graphite, carbon nanotubes and various kinds of metal particles. Some commercially available TiO_2 formulations are now able to use indoor ambient light for photocatalysis. A couple of examples are KRONOClean 7000 and Activa photocatalytic paint. The photocatalytic air purifier described in this guide uses the Activa paint.

But TiO_2 still seems to work most efficiently as a photocatalyst when it is used mostly in the form of a pure powder of anatase crystals irradiated by an ultraviolet light source. The light source can be something as simple as an ultraviolet fluorescent lamp or UV light emitting diodes. One of the problems with using such a light source however is that prolonged human exposure to such light can be dangerous. Any air purifier would have to be shielded so that very little light escapes if it is to be used in close proximity to people.

Are there any viable alternatives to TiO_2? Several other semiconductors have been studied for their photocatalytic properties. The most common ones are zinc oxide ZnO, tungsten trioxide WO_3 and tin oxide SnO_2. They all have band gap energies very similar to TiO_2 so they also require short wavelength light for photoactivation.

More important than the band gap energy is the actual energy of the lower edge of the conduction band and the upper edge of the valence band. These energies should be such that the oxidation and reduction processes discussed above can take place. TiO_2 is almost unique in this respect, allowing the oxidation of water and the reduction of oxygen and hydrogen ions. The only similar semiconductor is ZnO but it suffers from photocorrosion problems i.e. it is not as stable as TiO_2 under prolonged irradiation. So for the time being TiO_2 seems to be the most useful photocatalyst for air purification.

The details of photocatalytic process are still far from being well understood. One of the most important details involves the recombination rate. This is the rate at which photon excited electrons and holes recombine. It is the major limiting factor in the efficiency of photocatalysis. One of the major researchers in this area, Bunsho Ohtani, had this to say about the recombination rate problem:

"When a new photocatalyst shows increased photocatalytic reaction rates it is attributed to a reduction in the electron-hole recombination rate. Likewise a reduced reaction rate is attributed to an increase in electron-hole recombination rate. This is true despite the fact that the mechanism behind changes in recombination rates is never discussed and is presumably unknown." (ref. 35)

In another paper the same author states:

"It is surprising, at least for the author, that we know little about heterogeneous photocatalysis even though extensive studies have been carried out for more than forty years after the first-generation boom induced by the publication of a paper by Fujishima and Honda. What does this mean? There should be missing concepts, properties of photocatalysts or primary steps in photocatalysis that have not been noticed or understood but prevent progress of studies in the field of photocatalysis. ... The most important but unknown point must be e-h recombination. The only thing we know about the recombination in particulate photocatalysts is that it occurs to result in quantum efficiency smaller than 100%. As described above, we don't know where, when and how the recombination occurs." (ref. 34)

So there is much research still to be done in this area. There is the potential for new discoveries that could have an impact on not just cleaning up the environment but also on environmentally friendly energy production. The first application of TiO_2 as a photocatalyst was in the photolysis of water into hydrogen and oxygen gas which can then be used as fuel. This is essentially a form of solar energy storage. The work was done by Akira Fujishima, Kenichi Honda in 1972, see reference 1 in the bibliography. This paper launched the field of TiO_2 photocatalysis studies that continues to this day.

If you are interested in learning more about this field, see the references in the bibliography. There is no shortage of reviews of the field. Some of the better ones are listed in the bibliography.

How to Make a Photocatalytic Air Purifier

Creating the photocatalytic surface

Photocatalysis is basically a chemical reaction that takes place on a surface so let's start by creating the surface. This is the funnest part of the project. It begins by eating two cans of pineapple and includes some painting along the way, reminds me of kindergarten in Hawaii many, many years ago.

We take two cans of Great Value pineapple (net wt 20oz, 567gm), chunks, slices, crushed or tidbits, it doesn't matter, any pineapple will do. These cans currently cost $1.28 each from our local Walmart. The outer diameter of these cans is 3 9/32 inch = 3.2830 inch = 83.38 mm. The height is 4 59/128 inch = 4.4605 inch = 113.30 mm.

Figure 2: Great Value pineapple (net wt 20oz, 567gm) can

While the particular brand of pineapple is not important, as long as the can is about the same size, it is important that the can contain pineapple and not some other fruit. This is because pineapple cans do not have a plastic liner, rather they are lined with tin. The tin consumes the oxygen in the can, preventing the pineapple from oxidizing and turning brown [1].

Most food cans are lined with plastic. If you look inside both a tin lined pineapple can, and a can lined with plastic, you can see the difference. The pineapple can has a shiny metalic look, while the plastic lined one has a dull glazed look.

[1] https://extension.illinois.edu/blogs/live-well-eat-well/2022-02-16-lets-look-inside-can

plastic lined can   tin lined pineapple can

Figure 3: Plastic lined and tin lined can

If you can't find a pineapple can about the same size, there may be other fruit cans that are also lined with tin, which you could use. Fruits which are easily oxidized are stored in tin lined cans to prevent browning. If you can find applesauce in a can, that might also work, as long as the inside of the can looks like the inside of a pineapple can and not a plastic lined can.

The photocatalytic paint will stick onto the tin lined can well, better than it would to plastic. It just so happens that tin is also photocatalytic, although in the ultraviolet range, so it's a nice choice for the substrate of photocatalytic paint.

After consuming the contents of the two pineapple cans, remove the bottom of the cans in the same way you removed the top. Rinse out the cans well. We want to paint the inside of the cans with the photocatalytic paint. The cans don't have to be completely dry when you start painting them, since the paint we use is water based, but they shouldn't be excessively wet either because we want a uniform coating.

We've found that a 1 inch (2.54 cm) wide paintbrush is a good sized brush to use: not too small that it takes forever, not too big that you can't reach inside easily.

Figure 4: Activa interior white matte finish (Photo Deco) photocatalytic visible light paint

The paint we use is Activa interior white matte finish (Photo Deco) photocatalytic visible light paint. You can buy it directly from the company at
https://activacoating.com/product/white-matte-finish-photo-deco/
for $67.97 a gallon (3.785 liters) or from Amazon at
https://www.amazon.com/dp/B07L31V5SX
for the same price. If you don't need a gallon (you can

paint it on your indoor walls as well), you can buy just a pint (0.473 liters) for $25.99 at Amazon.

Open the can of paint, and stir it until the solids that were on the bottom are mixed in well. Paint the inside of the can until it is uniformly white. If you want, you can apply a second coat of paint after 6 to 8 hours, according to the instructions on the can. When finished, rinse out the brush with water.

With the inside of the two cans painted, and the bottoms cut out, as well as the tops, we want to connect the two cans together to make one long tube. To do this, we use a piece of aluminum tape. You can use other ways to connect the two cans, but the aluminum tape seals the connection fairly well, and the tape color matches the shiny metallic can surface, giving it a space-age kind of look. Aluminum tape is widely available (Home Depot, Lowes, Walmart, Harbor Freight, Amazon). One long strip of tape along the outer perimeter of where the two cans come together will connect them nicely.

Finally, we apply a couple rings of self-stick 1/4 inch (6.35 mm) Frost King vinyl foam tape to one end of the photocatalysis chamber we just made. This lets the chamber sit firmly on its base without rocking or vibrating due to the fan, and seals the interface with the base permitting the air to flow out where intended.

Figure 5: Completed photocatalysis chamber side view.

Mounting the fan and filter

Now we'll create the top portion of the photocatalytic air purifier that holds the fan and filter. The fan will push the air down through, first the filter, then the photocatalytic surface on the inside of the cans, before the air exits below the cans.

The mount from which the fan hangs and the filter sits atop is made of clear or white acrylic that is $1/8$ inch $=$ 0.125 inch $=$ 3.175 mm thick. Avoid black acrylic because we don't want much of the light inside the photocatalysis chamber to be absorbed by the acrylic. The white acrylic reflects the most light so, technically, it's the best, but the clear acrylic is OK, reflecting some of the light, while the rest produces a nice lamp effect at the top of the air purifier. You can make clear acrylic more white by buffing it with sandpaper.

We want to cut a ring out of the acrylic. The outer diameter of the ring is 4 inch $=$ 10.16 cm. The inner

Figure 6: Completed photocatalysis chamber top view.

diameter is 6.0 cm because we are using a 60 mm fan (6.0 cm = 2.362 inch ≈ 2 3/8 inch). We also need to drill four holes into the ring to mount the fan. These holes need to be 4 mm = 0.157 inch ≈ 5/32 inch in diameter, and arranged in a square pattern with 50 mm between the centers of the holes. See figure 7.

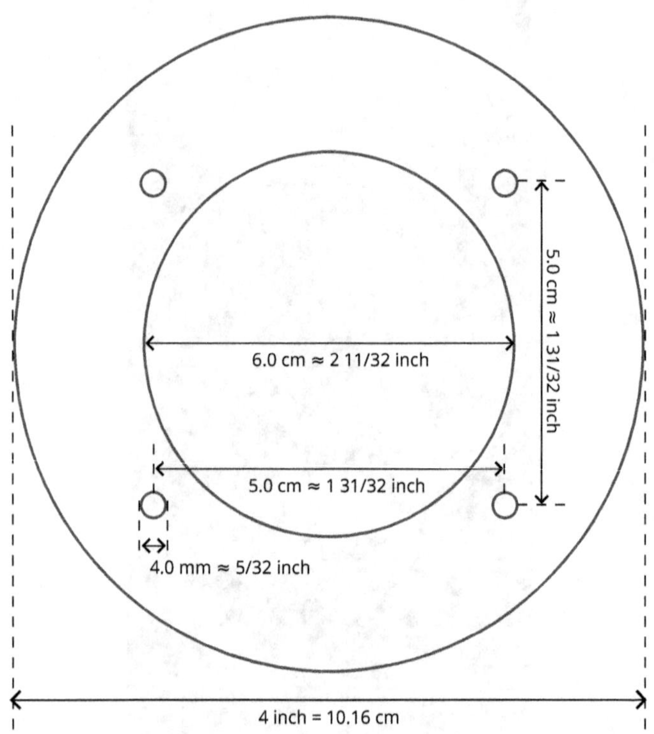

Figure 7: Acrylic ring for mounting fan and filter

How can you cut a ring out of 1/8 inch acrylic? Some options might be an adjustable hole cutter with a drill press, a band saw, a jig saw, a Dremel type of rotary tool, or an oscillating multi-tool. You might find it advantageous to cut out the inner ring first, as it can be easier to hold the acrylic that way when cutting.

To mount the fan you will need 4 machine screws that are no less than 1 inch $=$ 2.54 cm long and no wider in diameter than 4 mm $=$ 0.157 inch along the threads. An imperial screw size that works is #6-32 x 1 inch. These screws are 3.57 mm in diameter. A metric size that would work is M4-0.7 x 25 mm, which are 4 mm in diameter and 25 mm long.

When mounting the fan onto the acrylic ring, you want to make sure that the air flow exiting the fan is away from the ring. Do this by looking for an arrow on the side of the fan which indicates the direction of air flow. The screw heads should be on the upstream side of the fan because on that side we want to place a filter material to keep the particulate matter out of the photocatalysis chamber. The filter material we use is about 1/16 inch (1.59 mm) thick and has a MERV 13 particulate efficiency rating. Between the filter material and the fan, we place a screen that has a 1/4 inch (6.35 mm) spacing, to prevent the material from being sucked into the fan. This screen can be cut out to the right size with a small wire cutter. The screen is fastened to the acrylic ring with washers under the head of the machine screws. The washers are for #6 screws with 3/8 inch (9.5 mm) outer diameter and 5/32 inch (4 mm) inner diameter.

A few neodymium disk magnets of 5/16 inch (8 mm) diameter and 1/8 inch (3 mm) thickness are used to magnetically hold the filter material to the screen.

Figure 8: Fan mounted onto acrylic ring with screen below ring.

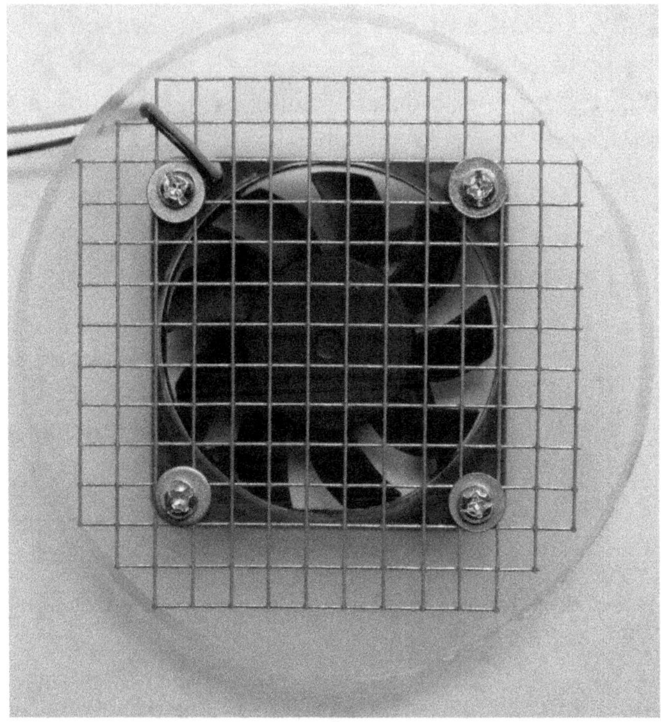

Figure 9: Fan mounted onto acrylic ring with screen above ring.

Setting up the light source

Our light source is a screw-in type of LED light bulb. The important parameter is that the bulb must be specified as 5000K or higher. This number indicates color temperature (K stands for Kelvin). A 5000K bulb has a blue tint to it, as opposed to a 3000K bulb which has a yellow tint. 5000K bulbs are also described as "daylight" bulbs, since their color is closer to sunlight than the 3000K bulbs. A few examples are shown below:

- Feit Electric Led 60W Replacement Day Light (6 in 1 Pack) $14.99 at Amazon.

- Great Value LED Light Bulb, 14 Watts (100W Equivalent) A19 General Purpose Lamp E26 Medium Base, Non-dimmable, Daylight, 4-Pack $10.97 at Walmart.

Both of the above listed bulbs are size A19 with an E26 bulb base.

The higher the color temperature number the better. We haven't seen screw-in type LED bulbs higher than 6000K.

The bulb socket we use to hold and wire up the bulb is: Leviton Plastic Keyless Lamp Holder $2.33 at Home Depot.

We want the light from the bulb to directly shine onto the photocatalytic surface, so we remove the plastic or glass covering that gives the bulb its familiar shape. We do this by placing the bulb in a bag then hitting the bulb with a hammer, trying not to damage the LED filament inside. If the bulb covering is glass, as ours was, it will shatter, and you can throw the broken pieces away. If it's plastic, you may need to cut around the base of the bulb with a sharp blade to remove the cover, being careful not to cut your hand, or damage the LED filament. Note that removing the bulb covering does expose the wires that power the LEDs, so be careful not to touch them when the lamp is powered up.

The lamp base needs to be connected to AC power. You can do this by using an old 2-prong appliance cord, wiring it up as shown in figure 10. If you don't have experience with AC power wiring, find someone who does, since this can be dangerous.

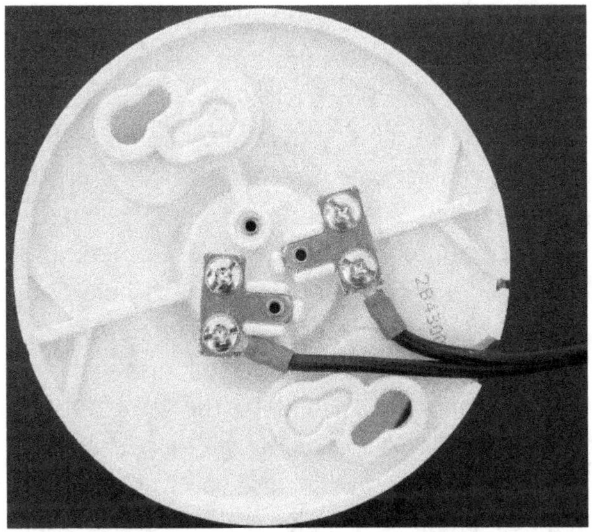

Figure 10: Wiring AC power to lamp base.

Coupling light to the photocatalysis chamber

To interface the light with the photocatalysis chamber we use a 3 inch (7.6 cm) PVC pipe coupler you can buy in Home Depot. Note that the outer diameter of this coupler is actually 4 inches (10.2 cm) and the inner diameter is 3.5 inches (8.9 cm). This allows a 3 inch (7.6 cm) PVC pipe, which actually has an outer diameter of 3.5 inches (8.9 cm), to fit into it.

The chamber slides into one end of the coupler. The end of the chamber that slides into the coupler should be the one that has the Frost King vinyl foam tape around its circumference. This should provide an airtight seal between the coupler and the chamber. The other end of the coupler has 16 holes drilled into it, spaced evenly around its circumference. The hole diameters are 3/4 inch (1.9 cm), with the bottom of each hole about 3/8 inch (9.5 mm) from the end of the coupler. It is this end of the coupler that sits on top of the lamp base. The coupler can be attached to the lamp base with a few pieces of scotch tape or hot glue. It should only be semipermanently attached in case you want to change the light source. The pipe coupler with the holes drilled into it is shown in figure 11, and the completed assembly is shown in figures 12 (top view) and 13 (side view).

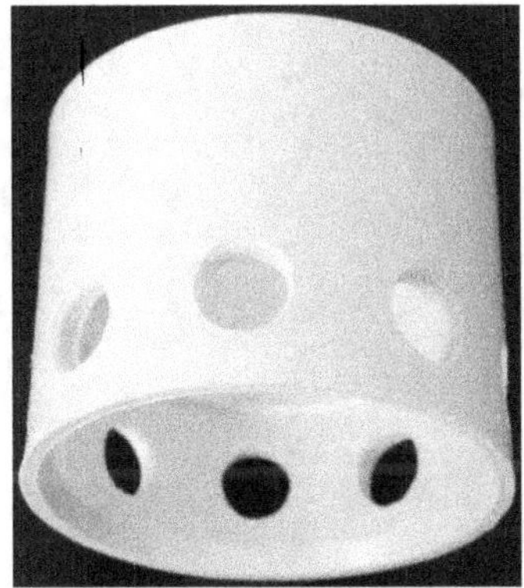

Figure 11: Pipe coupler with holes drilled into it.

Figure 12: Pipe coupler on lamp base (top view).

Figure 13: Pipe coupler on lamp base (side view).

The fan power supply

Any 12 volt power supply can be used to power the fan. In an early prototype we simply used an old 12 volt wall wart. For a more compact solution you can use the NOYITO AC to DC Precision Buck Power Supply Module AC 120V to 12V 500mA which we got at Amazon for $7.99. A picture of the module as well as the module connected to the fan is shown in figures 14 and 15, respectively.

Figure 14: Power supply module with input and output wires soldered on.

The AC side of the module has wires soldered onto it that will connect to the AC line powering the light beneath the lamp base as shown in figure 16.

The DC side of the module has plus ($+$) and minus (-) symbols on it. We solder wires onto that, which are then connected directly to the fan. Be sure to connect the plus wire from the module to the red wire on the fan, and the minus wire from the module to the black wire on the fan. This ensures the proper direction of airflow. The 12 volt supply should be housed in an enclosure to reduce

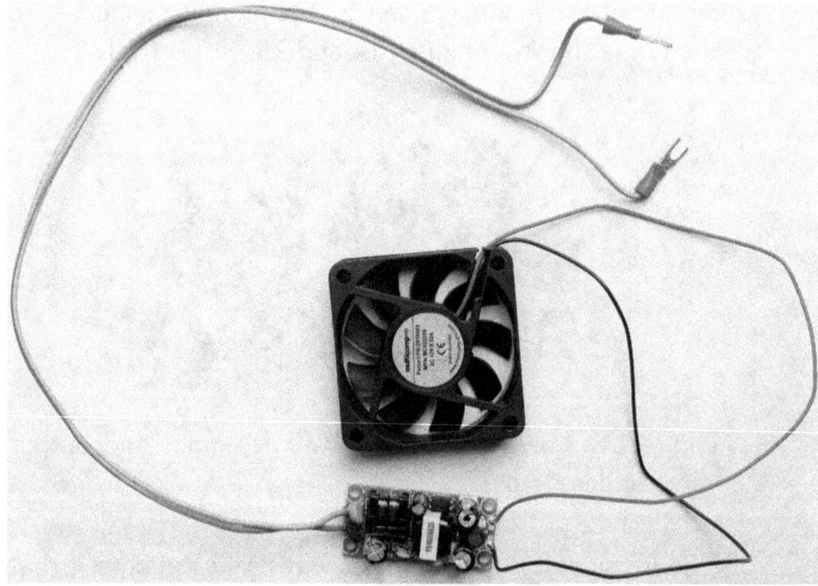

Figure 15: Power supply module connected to fan.

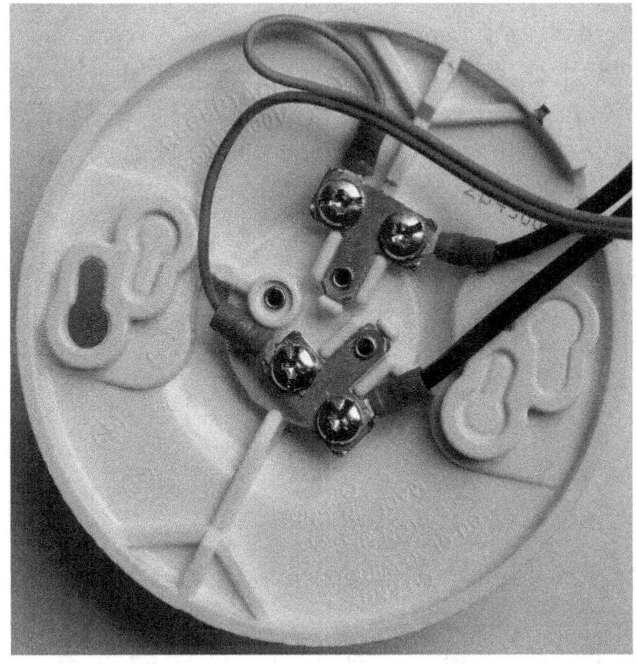

Figure 16: Lamp base wiring with AC line and power module connected.

shock hazard. This you can 3D print yourself, or use an enclosure you have readily available. Even an empty supplement bottle could do. The board's dimension are 2 inch x 1 inch x 5/8 inch (51 mm x 25.4 mm x 16 mm). Be sure to have some ventilation holes in the enclosure so the power supply module dissipates heat better.

After the wiring beneath the lamp base is completed you should place a cover over that to reduce shock hazard. One option for doing this is to cut a 4 inch (10.2 cm) diameter circle from a rubber packing sheet you can get at Home Depot, then attach it to the base with a little double sided carpet tape. This covering not only reduces shock hazard, but also reduces vibration caused by the fan, where the tabletop meets the lamp base of the air purifier.

Parts list

1. Two Great Value pineapple (net wt 20oz, 567gm) cans 2 x $1.28 = $2.56
   https://www.walmart.com/ip/10315437

2. Activa interior white matte finish (Photo Deco) photocatalytic visible light paint for $67.97 a gallon (3.785 liters) from the distributor at
   https://activacoating.com/product/white-matte-finish-photo-deco/
   or from Amazon at
   https://www.amazon.com/dp/B07L31V5SX
   for the same price, or buy just a pint (0.473 liters)

from Amazon for $25.99.

3. Aluminum tape. Widely available:
https://www.homedepot.com/p/317579705 $4.97
https://www.lowes.com/pd/3M-2-5-in-x-30-ft-Pipe-Wrap-Tape/1002947412 $6.98
https://www.walmart.com/ip/530697160 $2.63
https://www.harborfreight.com/96372.html $6.99
https://www.amazon.com/dp/B09BHWMSTD/ $6.99

4. Self-stick 1/4 inch (6.35 mm) Frost King vinyl foam tape
https://www.homedepot.com/p/300301768 $1.19

5. 60 mm computer fan
https://www.newark.com/multicomp-pro/mc002109/axial-fan-60mm-12vdc-18-84cfm/dp/45AC5492 $1.44

6. 1/8 inch (3.2 mm) thick white acrylic
https://www.canalplastic.com/products/7508-white-opaque-acrylic-sheet $3.50 for 6 x 12 inch (15 x 30 cm)
or
https://www.amazon.com/dp/B0818HYY8G/ $9.99 for 12 x 12 inch (30 x 30 cm)
or
https://www.michaels.com/product/mm-8x10-225-acrylic-10689048 $5.99 for 8 x 10 inch (20 x 25 cm), 2/25 inch (2 mm) thick (near enough to 1/8 inch to work)

7. Four machine screws with nuts:
#6-32 x 1 inch
https://www.homedepot.com/p/204274605 $1.38

or
M4-0.7 x 25 mm
https://www.lowes.com/pd/Hillman-4mm-0-7-x-25mm-Phillips-Drive-Machine-Screws-12-Count/999994838
$2.48

8. MERV 13 filter material (about 1/16 inch = 1.59 mm thick, 16 square feet = 1.49 square meters) https://www.amazon.com/dp/B087G5BY6G $17.99

9. Steel screen (mesh) with 1/4 inch (6.35 mm) spacing, 2 feet (61 cm) x 5 ft (152 cm) https://www.homedepot.com/p/205960850 $14.26

10. Washers for #6 screws with 3/8 inch (9.5 mm) outer diameter https://www.lowes.com/pd/Hillman-36-Count-x-3-8-in-Stainless-Steel-Standard-SAE-Flat-Washer/3816111 $2.98

11. Neodymium disk magnets of 5/16 inch (8 mm) diameter and 1/8 inch (3 mm) thickness https://www.harborfreight.com/10-piece-rare-earth-magnets-67488.html $2.99 for 10 pack
or
https://www.amazon.com/dp/B09TQQHQLP/ $5.99 for 50 pack

12. LED light bulb
Feit Electric Led 60W Replacement Day Light (6 pk)
https://www.amazon.com/dp/B07R7W6MBW/ $14.99
or

Great Value LED Light Bulb, 14 Watts (100W Equivalent) A19 (4 pk)
https://www.walmart.com/ip/53017078 $10.97

13. Bulb socket - Leviton Plastic Keyless Lamp Holder
https://www.homedepot.com/p/100356849 $2.33

14. 3 inch (7.6 cm) PVC pipe coupler (outer diam = 4 inches (10.2 cm), inner diam = 3.5 inches (8.9 cm))
https://www.homedepot.com/p/203390807 $3.84

15. NOYITO AC to DC Precision Buck Power Supply Module AC 120V to 12V 500mA
https://www.amazon.com/dp/B07FNJZ1PR/ $7.99

16. Rubber packing sheet, 6 x 6 inch (15.2 cm)
https://www.homedepot.com/p/203193498 $6.25

17. Double sided carpet tape
https://www.walmart.com/ip/17510811 $6.84

The total minimum cost (choosing lowest price options) without shipping cost for this photocatalytic air purifier is $115.13.

Some of the above items are consumed in quite small quantities for making a single photocatalytic air purifier, resulting in a total cost that is much higher than it could be if you could buy in smaller quantities. As a service, we're offering a kit to build a single photocatalytic air purifier. See this guide's webpage for details:

https://www.abrazol.com/guides/photocat1/

Bibliography

1. Electrochemical Photolysis of Water at a Semiconductor Electrode
   Akira Fujishima, Kenichi Honda
   Nature volume 238, p. 37–38 (1972)

2. Environmental Applications of Semiconductor Photocatalysis
   Michael R. Hoffmann, Scot T. Martin, Wonyong Choi, Detlef W. Bahnemann
   Chem. Rev. 1995, 95, 69-96

3. Photoinduced reactivity of titanium dioxide
   O. Carp, C.L. Huisman, A. Reller
   Progress in Solid State Chemistry 32 (2004) 33–177

4. Understanding TiO2 Photocatalysis: Mechanisms and Materials
   Jenny Schneider, Masaya Matsuoka, Masato Takeuchi, Jinlong Zhang, Yu Horiuchi, Masakazu Anpo, Detlef W. Bahnemann
   Chem. Rev. 2014, 114, 9919-9986

5. Removal of indoor volatile organic compounds via

photocatalytic oxidation: a short review and prospect
Yu Huang, Steven Sai Hang Ho, Yanfeng Lu, Ruiyuan
Niu, Lifeng Xu, Junji Cao, Shuncheng Lee
Molecules, Vol 21, No 1, Jan 4, 2016, 20p

6. A brief survey of the practicality of using photo-catalysis to purify the ambient air (indoors or out-doors) or air effluents
Pierre Pichat
Applied catalysis B: environmental, Vol 245, No 9, May 15, 2019, p770-776

7. The viability of photocatalysis for air purification
Stephen O. Hay, Timothy Obee, Zhu Luo, Ting Jiang, Yongtao Meng
Molecules 2015, 20, 1319-1356

8. Photocatalytic air purification mimicking the self-cleaning process of the atmosphere
Fei He, Woojung Jeon, Wonyong Choi
Nature Communications 2021, 12:2528

9. Heterogeneous photocatalysis: recent advances and applications
Alex Omo Ibhadon, Paul Fitzpatrick
Catalysts 2013, 3, 189-218

10. Advances and challenges of photocatalytic technol-ogy for air purification
Qin Geng, Hong Wang, Ruiming Chen, Lvcun Chen, Kanglu Li, Fan Dong
Natl. Sci. Open, 2022, Vol. 1, 20220025

11. A brief overview of TiO2 photocatalyst for organic dye remediation: case study of reaction mechanisms

involved in Ce-TiO2 photocatlysts system
Milind Pawar, S. Topcu Sendogdular, Perena Gouma
Journal of Nanomaterials, Vol. 2018, 5953609

12. Photocatalytic purification of volatile organic compounds in indoor air: a literature review
Jinhan Mo, Yinping Zhang, Qiujian Xu, Jennifer Joaquin Lamson, Rongyi Zhao
Atmospheric Environment 43 (2009 2229-2246

13. Removal of indoor volatile organic compounds via photocatalytic oxidation: a short review and prospect
Yu Huang, Steven Sai Hang Ho, Yanfeng Lu, Ruiyuan Niu
Molecules 2016, 21, 56

14. Photocatalytic air cleaners and materials technologies - abilities and limitations
Lexuan Zhong, Fariborz Haghighat
Building and Environment xxx (2015) 1-13

15. TiO2 photocatalyst for removal of volatile organic compounds in gas phase
Zahra Shayegan, Chang-Seo Lee, Fariborz Haghighat
Chemical Engineering Journal, Vol. 334, 15 February 2018, Pages 2408-2439

16. An overview of recent developments in improving the photocatalytic activity of TiO2 based materials for treatment of indoor air and bacterial inactivation
Achraf Amir Assadi, Oussama Baaloudj, Lotfi Khezami, Naoufel Ben Hamadi, Lotfi Mouni, Aymen Amine Assadi, Achraf Ghorbal
Materials 2023, 16, 2246

17. Efficiencies evaluation of photocatalytic paints under indoor and outdoor air coditions
Federico Salvadores, Martin Reli, Orlando M. Alfano, Kamila Koci, Maria de los Milagros Ballari
Frontiers in Chemistry, October 2020, V8, Article 551710

18. A review on the visible light active titanium dioxide photocatalysts for environmental applications
Miguel Pelaez, Nicholas T. Nolan, Suresh C. Pillai, et al
Applied Catalysis B: Environmental 125 (2012) 331-349

19. Photocatalytic decomposition of acetaldehyde on different TiO2 based material: a review
Beata Tryba, Piotr Rychtowski, Agata Markowska-Szczupak, Jacek Przepiorski
Catalysts 2020, 10, 1464

20. Quantitative XRD (X-ray diffraction) characterisation and gas phase photocatalytic activity testing for visible light (indoor applications of KRONOClean 7000
D. M. Tobaldi, M. P. Seabra, G. Otero-Irureta, et. al.
RSC Advances, 2015, 5, 102911

21. Photocatalytic construction and building materials: From fundamentals to applications
Jun Chen, Chi-sun Poon
Building and Environment 44 (2009) 1899–1906

22. Photocatalytic disinfection using titanium dioxide: spectrum and mechanism of antimicrobial activity
Howard A. Foster, Iram B. Ditta, Sajnu Varghese, Alex Steele
Appl Microbiol Biotechnol (2011) 90:1847–1868

23. Titanium dioxide photocatalysis
Akira Fujishima, Tata N. Rao, Donald A. Tryk
Journal of Photochemistry and Photobiology C: Photochemistry Reviews 1 (2000) 1–21

24. Titanium dioxide photocatalysis: fundamentals and application on photoinactivation
Pedro Magalhaes, Luísa Andrade, Olga C. Nunes and Adelio Mendes
Rev. Adv. Mater. Sci. 51 (2017) 91-129

25. Titanium Dioxide Photocatalysis in Atmospheric Chemistry
Haihan Chen, Charith E. Nanayakkara, Vicki H. Grassian
Chem. Rev. 2012, 112, 5919-5948

26. Titanium dioxide photocatalysis: present situation and future approaches
Akira Fujishima, Xintong Zhang
C. R. Chimie 9 (2006) 750–760

27. Development of alternative photocatalysts to TiO2 : Challenges and opportunities
Maria D. Hernandez-Alonso, Fernando Fresno, Silvia Suarez, Juan M. Coronado
Energy Environ. Sci., 2009, 2, 1231–1257

28. Semiconductor Composites: Strategies for Enhancing Charge Carrier Separation to Improve Photocatalytic Activity
Roland Marschall
Adv. Funct. Mater. 2014, 24, 2421–2440

29. Titanium Dioxide: From Engineering to Applications
Xiaolan Kang, Sihang Liu, Zideng Dai, Yunping He, Xuezhi Song, Zhenquan Tan
Catalysts 2019, 9, 191

30. A Promising Technological Approach to Improve Indoor Air Quality
Thomas Maggos, Vassilios Binas, Vasileios Siaperas, Antypas Terzopoulos, Panagiotis Panagopoulos, George Kiriakidis
Appl. Sci. 2019, 9, 4837

31. Photocatalysis on TiO2 Surfaces: Principles, Mechanisms, and Selected Results
Amy L. Linsebigler, Guangquan Lu, and John T. Yates, Jr.
Chem. Rev. 1995, 95, 735-758

32. Photocatalytic generation of gas phase reactive oxygen species from adsorbed water: remote action and electrochemical detection
Xinwei Sun, Kaiqi Xu, Athanasios Chatzitakis, Truls Norby
Journal of Environmental Chemical Engineering, Vol 9, 14809, 2021

33. Photocatalytic Decomposition of Air Pollutants Using Electrodeposited Photocatalysts on Stainless Steel
Andreas Hanel, Marcin Janczarek, Marek Lieder, Jan Hupka
Pol. J. Environ. Stud. Vol. 28, No. 3 (2019), 1157-1164

34. Revisiting the fundamental physical chemistry in heterogenous photocatalysis: its thermodynamics and kinetics
Bunsho Ohtani
Physical Chemistry Chemical Physics, 2014, Vol 16, No 5, p1788-1797

35. Titania photocatalysis beyond recombination: a critical review
Bunsho Ohtani
Catalysts 2013, 3, 942-953

36. Fully quantitative X-ray characterisation of Evonik Aeroxide TiO2 P25
D.M. Tobaldi, R.C. Pullar, M.P. Seabra, J.A. Labrincha
Materials Letters 122 (2014) 345–347

37. What is Degussa (Evonik) P25? Crystalline composition analysis, reconstruction from isolated pure particles and photocatalytic activity test
B. Ohtani, O. O. Prieto-Mahaney, D. Li, R. Abe
Journal of Photochemistry and Photobiology A : Chemistry, 216(2-3), 179-182

About the Authors

Stefan Hollos and **J. Richard Hollos** are brothers and business partners at Exstrom Laboratories LLC (www.exstrom.com) in Longmont, Colorado. They are physicists and electrical engineers by training, and enjoy anything related to math, physics, engineering and computing. In addition, they enjoy creating music and visual art, and being in the great outdoors. They are the authors of the following books:

- **Creating Noise, second edition**

- **Engineer's Notebook on Inductor and Transformer Circuits: Problems, Solutions and Simulations**

- **The Enigma of the Crookes Radiometer**

- **Passive Butterworth Filter Cookbook**

- **Nell: An SVG Drawing Language**

- **Coin Tossing: The Hydrogen Atom of Probability**

- **Creating Melodies**

- **Hexagonal Tilings and Patterns**

- **Combinatorics II Problems and Solutions: Counting Patterns**

- **Information Theory: A Concise Introduction**

- **Recursive Digital Filters: A Concise Guide**

- **Art of Pi**

- **Creating Noise**

- **Art of the Golden Ratio**

- **Creating Rhythms**

- **Pattern Generation for Computational Art**

- **Finite Automata and Regular Expressions: Problems and Solutions**

- **Probability Problems and Solutions**

- **Combinatorics Problems and Solutions**

- **The Coin Toss: Probabilities and Patterns**

- **Pairs Trading: A Bayesian Example**

- **Simple Trading Strategies That Work**

- **Bet Smart: The Kelly System for Gambling and Investing**

- **Signals from the Subatomic World: How to Build a Proton Precession Magnetometer**

More information on all these books can be found at the website of Abrazol Publishing
www.abrazol.com
where you can also sign up for the Abrazol newsletter.

Thank You

Thank you for buying this guide.

If you'd like to receive news about this guide and others published by Abrazol Publishing, just go to

http://www.abrazol.com/

and sign up for our newsletter.

9 798852 420602